HEATHER'S Amazing Discovery

A TRUE STORY OF PALEONTOLOGY

BY Deborah Griffiths

ILLUSTRATED BY Robert Lundquist AND Leah Pipe

WITH PHOTOGRAPHS BY Paul Bailey

All proceeds from this and any future
editions are dedicated to an endowment for
the Courtenay and District Museum

Copyright © 2013 by: Deborah Griffiths
All rights reserved.

ISBN: 0-9696-6120-7
ISBN-13: 9780969661207

*This book is dedicated to
the Trask family, and to the many
enthusiastic volunteers who
have helped bring the story of
Mike and Heather's discovery
to people of all ages...*

*And also, with love,
to Danby and Scott*

This is the true story of an amazing discovery made by a girl named Heather. Heather lives on Vancouver Island on the coast of British Columbia.

When Heather awoke on the morning of her discovery, she had no idea that what she and her father would find that day would become known around the world.

Heather's discovery began like this....

One sunny morning, Heather and her father, Mike, decided to spend the day looking for fossils on the nearby Puntledge River. They packed up things they would need for their hike, and headed from home with their knapsacks full. They had hopes of finding something special that day.

Heather and Mike walked along the trail toward the site where they knew they might have luck finding fossils. The winding trail followed the river. Tall fir, cedar and hemlock trees shaded them as they walked in the lush forest.

Heather enjoyed fossil hunting with her dad.

On their first trip to the river, he had explained that fossils are the remains of plants and animals that lived many, many years ago.

Over the years the plants and animals had been covered by layers of sand and mud and had gradually turned to stone.

In the recent past, the force of the river had worn through these layers of stone to expose fossils from an 80 million year old ocean floor. Each trip they made to the river was a journey back in time.

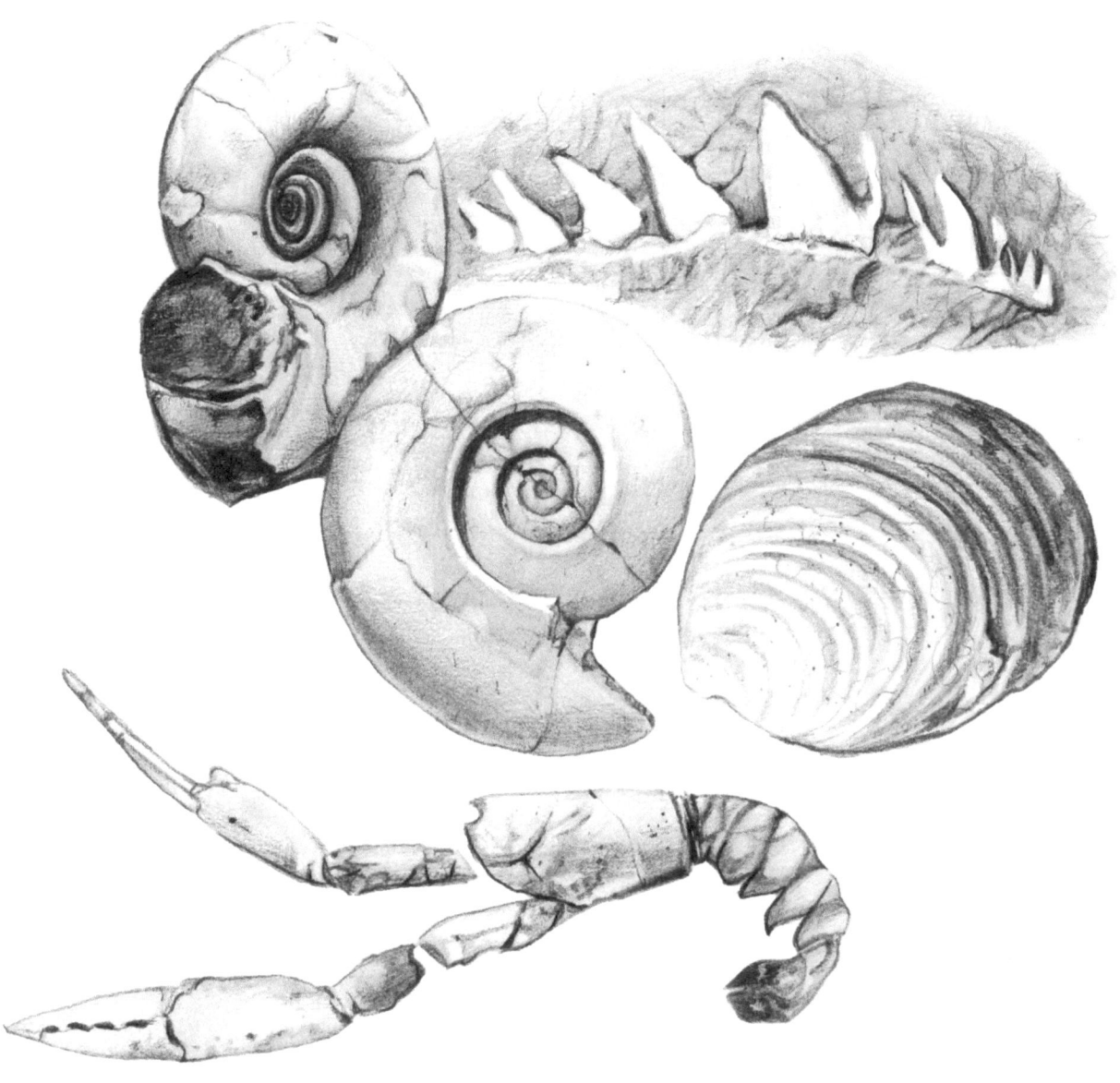

Some of the fossils they had found were ammonites, lobsters, sharks' teeth and shells. They were the remains of animals that lived in the ocean at the same time dinosaurs had lived on land.

They finally reached the river, and Heather began to walk along the bank and search the ground for fossils.

She began to daydream about how exciting it would be to discover…

Something larger and more mysterious than anything they had found yet,

Something ancient and extraordinary,

Something that lived in the ocean at the same time Tyrannosaurus rex, Triceratops and Pteranodon lived on land.

She wondered what it would be like to see these large animals living and breathing.

Suddenly Heather's daydreaming was interrupted by something she noticed at her feet. It was a rock as big as both her hands, standing out from the flat bed of shale. It looked different.

She kneeled down to take a closer look. It was the same grey colour as the shale, but smooth and round.

"Dad! Dad!" she called to her father, who was downriver. "Come have a look at this!"

Mike called, "What did you find, Heather?"

"I don't know," she said. "This looks different from anything we've found before."

Mike knelt to have a closer look.

He took a hammer and chisel out of his pack and carefully began removing the shale surrounding the stone.

Mike was very quiet as he worked. Heather had the feeling that this could be important. She wondered if they had found something that lived in the age of dinosaurs.

Finally, when she thought she couldn't wait any longer, Mike said, "Heather, I think we have a large bone. It could be a bone from a very large sea animal!"

There was more stone to be uncovered, and Heather waited impatiently as Mike worked. She listened to the hammer—clink, clink, clinking against the chisel—in tune with the sound of the river flowing over rocks and boulders.

As Mike finished with the stone he discovered another one next to it. Heather watched in amazement as he uncovered another…and another…and another bone! This was incredible!

They were all the same shape and size as the first one. For the next few hours Mike chiselled the rock surface while Heather helped clear the shale chips away.

By the end of the afternoon they had uncovered twelve large bones lying next to each other in a row. It was certainly part of the spine of a large animal.

Tired from digging, Heather and Mike decided to return to the site in the morning to see if they could find more bones.

That night, Heather and Mike searched through books to delve further into the mystery of life 80 million years ago.

With each page that Heather turned she learned more about the fantastical creatures of long ago.

Tomorrow seemed like a year away as she tried to sleep that night. She couldn't wait to return to the river in the morning and see the bones again.

But tired from her river hike, it wasn't long before she drifted off to sleep and began to dream....

19

What a dream! Heather was soaring through the air with animals from another time. Swimming below her were mighty beasts of the ocean.

She dove beneath the water's surface and entered an ancient ocean world.

There was a turtle, larger than Heather, that paddled through the water capturing fish in its sharp beak.

Even larger was the mosasaur. It looked like a giant sea crocodile, and whipped through the water searching for fish and ammonites to eat.

The elasmosaur was the largest of all. It used its long powerful neck and paddles to swim through the water, hunting for fish to swallow whole.

The next morning, Heather awoke and remembered her dream. Were the bones part of one of the sea creatures she had dreamed about? Could it be they had uncovered something so rare?

They packed up for another day at the river, and returned to the site of the discovery. Mike photographed the bones.

In a notebook he wrote down the location and described what the bones looked like. This would help scientists identify the animal.

Mike and Heather continued to dig for more bones, and by late that day they had uncovered eleven more. Now they had twenty-three in all. Heather was certain about one thing—this animal was incredibly long!

After studying the bones for several days, Mike decided to send some of them to a paleontologist to have them identified. He explained to Heather that a paleontologist is a scientist who studies ancient life.

When the paleontologist finished studying the bones, she returned them with a letter explaining that they had found the remains of an 80 million year old elasmosaur, the long-necked swimmer of Heather's dream.

Heather would never forget her amazing discovery.

The days she spent on the river with her father were always special—but the day they found the elasmosaur was one to remember forever!

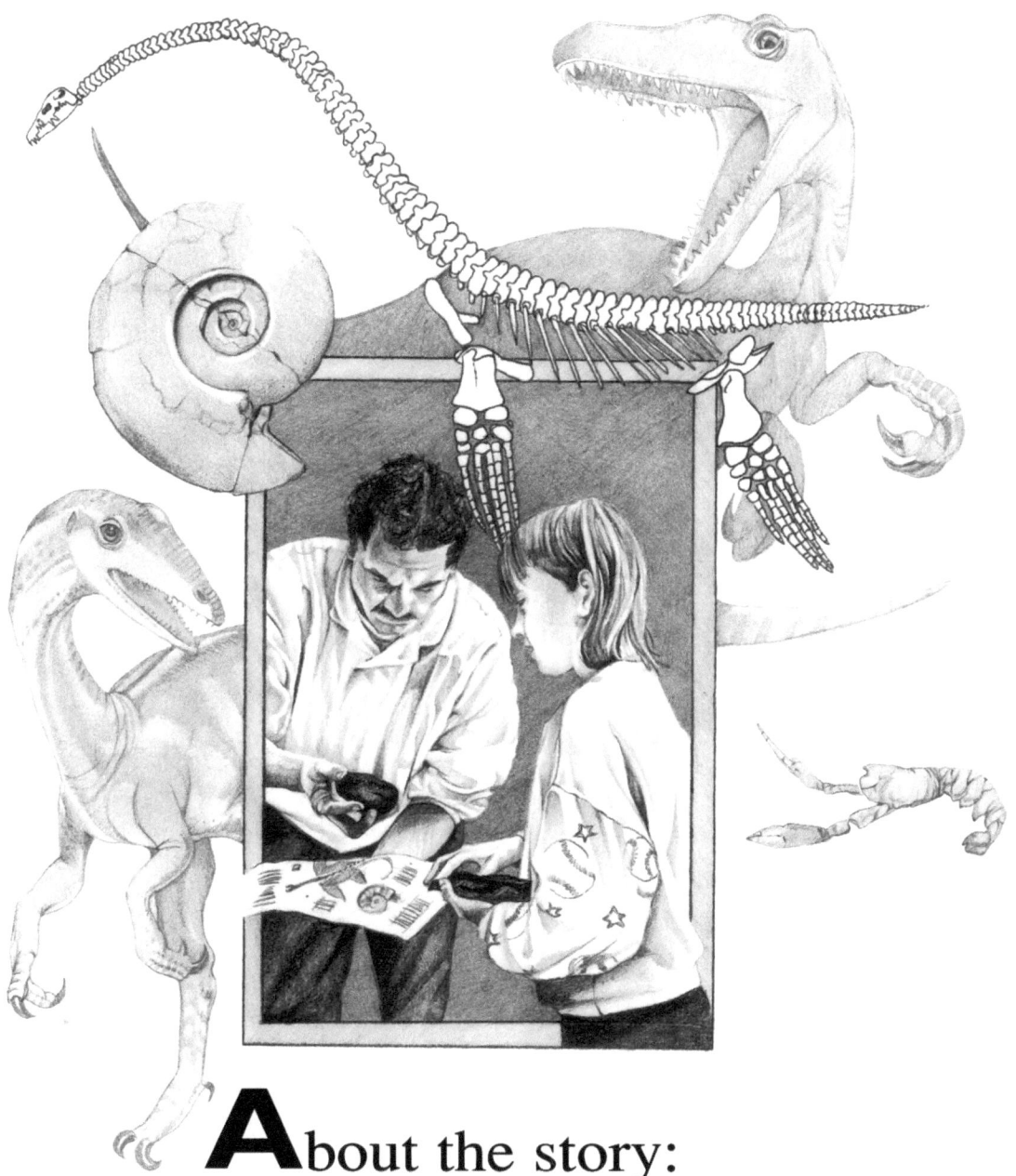

About the story:

Ever since the Trasks' amazing discovery in 1988, thousands of children and adults have learned about and participated in Courtenay Museum's paleontology programs, and discovered for themselves Vancouver Island's oceanic prehistory.

A large group of local volunteers led by Mike Trask and paleontologist Rolf Ludvigsen excavated the elasmosaur in 1991, and the reconstructed skeleton of this massive marine reptile is now hanging in the museum's galleries.

About the museum and paleontology in B.C.

In May 1984 Richard Boldt and his son Aaron found the remains of a marine turtle on the Puntledge River, in Courtenay, British Columbia, midway up Vancouver Island. The turtle was the first recorded Cretaceous marine reptile in B.C. Although the find was significant, it remained relatively unnoticed by the public until Mike and Heather Trask's discovery of the elasmosaur.

The Puntledge elasmosaur was identified by Dr. Betsy Nicholls in January 1989. It was the first elasmosaur west of the Canadian Rockies and the largest marine reptile ever found in the province.

The reconstructed elasmosaur, completed in 1995 by Prehistoric Animal Structures, is the culmination of a huge community effort since Heather's find

In the spring of 1989 the site was declared a protected heritage site. Then, in 1991, Courtenay Museum enlisted over forty volunteers, including a number of young people, to assist with the excavation of the rest of the elasmosaur. By the end of an arduous three-month quarry, the group had recovered over three-quarters of a complete elasmosaur.

Paleontologist Rolf Ludvigsen oversaw the elasmosaur excavation, and has watched interest in his profession soar

The exhibition OUR ANCIENT ISLAND was created to display the elasmosaur, along with specimens representative of a Cretaceous marine environment. The original bones were sent to Alberta to be cleaned and cast by Prehistoric Animal Structures, and a model of the fourteen-metre-long giant was mounted in the museum's exhibit.

Since the discovery, many other significant finds have been made locally. Two mosasaurs (a ferocious sea lizard) were found on the Puntledge River (by Zanbilowicz, Boldt and O'bear), and a second elasmosaur was found by Mike Trask and Lynn Roux on the Trent River in 1992. As well, Matthew Morin and his father, Joe, found the tooth of a theropod dinosaur on the Trent in 1994.

The Puntledge elasmosaur and related Cretaceous fossil material have received attention from visitors and researchers around the globe. Public interest in paleontology ignited by this once-in-a-lifetime discovery continues to grow. Young "Heathers" everywhere are catching a glimpse of the wonders of ancient life, and enriching their own lives through the experience.

Volunteer excavations are just one way the museum has brought the public into the process of examining ancient life through fossils

The excavations, presentations and other work of the Courtenay Museum are made possible by strong community support.

Thanks to everyone who has been involved with Heather's amazing discovery and its continuing preservation.

www.ingramcontent.com/pod-product-compliance
Lightning Source LLC
Chambersburg PA
CBHW041227040426
42444CB00002B/79